PATTERNS IN NUMBERS AND SHAPES

USING ALGEBRAIC THINKING

W9-DHF-136

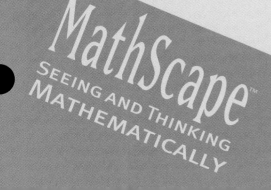

MathScape
SEEING AND THINKING
MATHEMATICALLY

You will practice looking for patterns in different places—in drawings and numbers, sometimes in a story. You will even make up some patterns of your own. By making tables of the data you find, you will discover ways to extend the patterns to very large sizes.

How does math relate to patterns?

PATTERNS IN NUMBERS AND SHAPES

PHASE**TWO**
Describing Patterns Using
Variables and Expressions

In this phase you look for
patterns in letters that grow and
chocolates in a box. You will
begin to use the language of
algebra by giving the rules for
these patterns using variables
and expressions. You will
practice using expressions to
compare different ways of
describing a pattern to see if
they give the same result.

PHASE**THREE**
Describing Patterns Using
Graphs

In this phase you will plot
points in all parts of the
coordinate grid to make a
mystery drawing for your
partner. You will turn number
rules into graphs and look at
the patterns they make. By
comparing graphs of some
teenagers' wages for summer
jobs, you will decide who has
the better pay rate.

PHASE**FOUR**
Finding and Extending Patterns

After you have learned some
ways to describe the rules for
patterns, this phase gives you a
chance to try out your skills in
new situations. You look for
patterns in a story about a
sneaky sheep and in animal
pictures that grow. You analyze
three different patterns for
inheriting some money in order
to give some good advice.

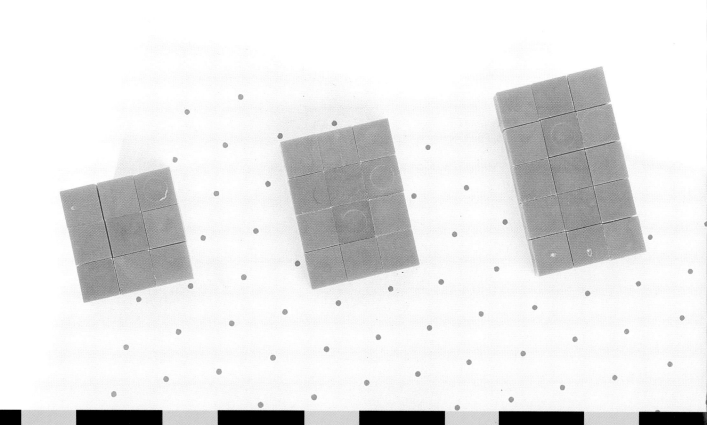

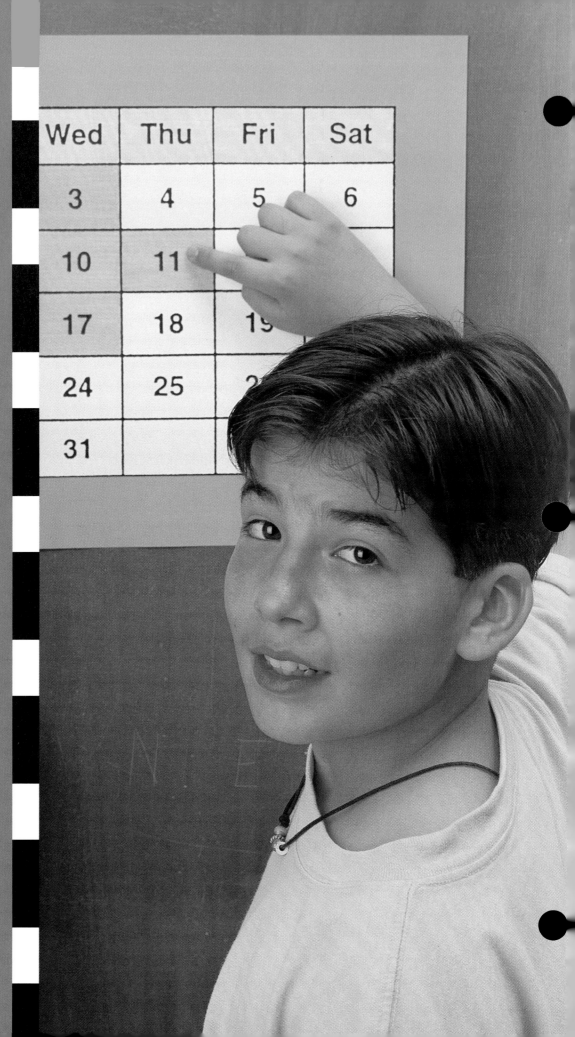

PHASE ONE

In this phase you will look for patterns in three different situations. By making tables of the data, you can see how the patterns develop and discover how to extend them to greater size. Working with patterns in numbers, shapes, and a story helps you to develop math skills that carry over into algebra problems. You will begin to be able to work out a rule that will apply to all cases of a situation, and use your rule to solve problems.

Describing Patterns Using Tables

WHAT'S THE MATH?

Investigations in this section focus on:

PATTERN SEEKING

- Developing skills in looking for patterns in new situations
- Writing rules for patterns
- Extending pattern rules to apply to all cases

RECORDING DATA

- Making tables of data to describe patterns

NUMBER

- Exploring number relationships to look for patterns

MathScape Online
mathscape1.com/self_check_quiz

1 Calendar Tricks

Finding a pattern can help you solve problems in surprising ways. In this activity, you will think about patterns as you test whether tricks using the numbers on a calendar will always be true. Then you will invent and test your own tricks.

Look for a Pattern

How can you know whether number patterns on a grid will always be true?

Paul invented three tricks for a block of four numbers. Find which of Paul's tricks are true for every possible two-by-two block of numbers on the calendar.

- Which tricks always worked? Which did not? How did you find out?

- For any tricks that did not always work, how can you revise them so that they do always work?

Paul's Box of Tricks

Sun	Mon	Tue	Wed	Thurs	Fri	Sat
1	2	3	4	5	6	7
8	9	10	11	12	13	14
15	16	17	18	19	20	21
22	23	24	25	26	27	28
29	30	31				

Trick One: The sums of opposite pairs of numbers will be equal. For example: $2 + 10 = 3 + 9$.

Trick Two: If you add all four numbers, the sum will always be evenly divisible by 8. For example:
$2 + 3 + 9 + 10 = 24; 24 \div 8 = 3$.

Trick Three: If you multiply opposite pairs of numbers, the two answers will always differ by 7. For example:
$2 \times 10 = 20; 3 \times 9 = 27; 27 - 20 = 7$.

Invent Your Own Tricks

How can you use patterns to invent tricks that will always work?

The calendar on this page shows sequences of three numbers going diagonally to the right, shaded blue, and to the left, outlined in red. Can you make up tricks about diagonals of three numbers like these?

Make up at least two tricks for diagonals of three numbers like the numbers shaded blue in the example.

- Will your tricks be true for every diagonal of three numbers? How do you know?

- Which of your tricks will still be true if the diagonal goes in the opposite direction like the numbers in the red outline?

Make up at least two more tricks using your own shapes. Your shapes should be different from the ones shown so far.

- Are your tricks always true, no matter where on the calendar you put your shape?

- How do you know your tricks will always be true?

Write About Finding Patterns

Think about the patterns you explored and the tricks you invented in this lesson.

- How did you check whether your tricks will work everywhere on the calendar?

- What suggestions would you give to help another student find patterns?

Patterns on a Slant

Sun	Mon	Tue	Wed	Thurs	Fri	Sat
1	2	3	4	5	6	7
8	9	10	11	12	13	14
15	16	17	18	19	20	21
22	23	24	25	26	27	28
29	30	31				

hot **words** | pattern

Homework
page 354

2 Painting Faces

EXPLORING
PATTERNS BY
ORGANIZING DATA

Here is a problem about painting all sides of a three-dimensional shape. Sometimes using objects is helpful in solving problems like this. You can record the data you get when you solve for shorter lengths and organize it into a table to help you find a rule for any length.

Make a Table of the Data

How can you set up a table to record data about a pattern?

A company that makes colored rods uses a paint stamping machine to color the rods. The stamp paints exactly one square of area at a time. Every outside face of each rod has to be painted, so this length 2 rod would need 10 stamps of paint.

How many stamps would you need to paint rods from lengths 1 to 10? Record your answers in a table and look for a pattern.

Length 2 rod;
each end equals 1 square.

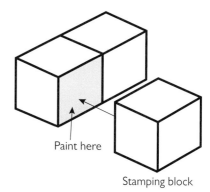

Paint here

Stamping block

How to Organize Your Data in a Table

1. Use the first column to show the data you begin with. Write the numbers in order from least to greatest. In this example, the lengths of the rods go in the first column.

2. Use the second column to show numbers that give information about the first sequence. Here, the numbers of stamps needed to paint each length of rod go in the second column.

Length of Rod	Stamps Needed
1	
2	
3	

Extend the Pattern

Look at the table you made. Do you see a pattern in the data in the table?

1 Use what you learned from the table. Find the number of paint stamps you need to paint rods of lengths 25 and 66.

2 Write a rule you could use to extend the pattern to any length of rod.

Can you write a rule you could use to extend the pattern?

Now look again at your table and your rule. Decide how you could use the number of stamps to find the size of the rod. Think about the operations you used in your rule.

3 What if it takes 86 stamps to paint the rod? How long is the rod?

4 If it takes 286 stamps to paint the rod, how long is the rod?

5 Write a rule you could use to find the length of any rod if you know the number of stamps needed to paint it.

Write About Making a Table

Think about how you used the table you made to solve problems about painting the rods.

- How did making a table help you to find the pattern?

- Describe how far you think you would need to extend a table to be sure of a pattern.

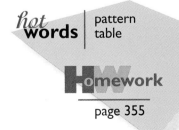

hot**words** | pattern
table

Homework
page 355

3 Crossing the River

Examining a pattern can help you develop a general rule that applies to any stage of the pattern. In this investigation you will look for a pattern to solve the problem of getting a group of hikers across a river using one small boat.

Find a Rule for Any Number

How can finding a pattern help solve for all cases?

Think carefully about how the hikers could cross the river using just one boat. It may be helpful to act it out or use a diagram to solve the problem. As you work, make a table showing how many trips it takes for 1 to 5 adults and 2 children to cross. Look for the pattern, then use it to find how many trips are required for the other groups to cross to the other side.

1 How many one-way trips does it take for the entire group of 8 adults and 2 children to cross the river? Tell how you found your answer.

2 How many trips in all for 6 adults and 2 children?

3 15 adults and 2 children?

4 23 adults and 2 children?

5 100 adults and 2 children?

Tell how you would find the number of one-way trips needed for any number of adults and two children to cross the river.
(Everyone can row the boat.)

Ten Hikers—One Boat

A group of 8 adults and 2 children needs to cross a river. They have a small boat that can hold either:

1 adult　　or　　1 child　　or　　2 children

Use Your Method in Another Way

Use the pattern to find the number of adults who need to cross the river for each case.

1 It takes 13 trips to get all of the adults and the 2 children across the river.

2 It takes 41 trips to get all of the adults and the 2 children across the river.

3 It takes 57 trips to get all of the adults and the 2 children across the river.

How can you work backward from what you know?

Tell How You Look for Patterns

Write a friend a letter telling how you look for patterns. Give examples from the patterns you have investigated so far. Answers to the following questions will help you write your letter.

- How can a table help you discover and describe a pattern?

- What other tools are helpful?

- How does finding a pattern help you solve problems?

hot words | pattern table

page 356

PHASE TWO

As you look for patterns in this phase, you will think about how you can describe their rules in a way that applies to all situations. You will begin to use the language of algebra by writing rules using variables and expressions. You will compare your rules to those of other students to see if they give the same result. Some of the patterns change in more than one way and you will need to find a way to express that in your rule.

Describing Patterns Using Variables and Expressions

WHAT'S THE MATH?

Investigations in this section focus on:

PATTERN SEEKING

- Looking for patterns in numbers and shapes
- Examining patterns with two variables

ALGEBRA

- Using variables and expressions

EQUIVALENCE

- Exploring equivalence of expressions

MathScape Online
mathscape1.com/self_check_quiz

4 Letter Perfect

Using variables and expressions gives you a shorthand way to describe a pattern. These tile letters grow according to different patterns. You will explore how to write a rule to predict the number of tiles needed to make letters of any size.

Find a Rule That Fits Every Case

How can a pattern help you predict the number of tiles used for any size?

Find a rule that will tell how many tiles it takes to build any size of the letter *I*.

Size 1   Size 2   Size 3

1 Look for a pattern. Describe it clearly with words.

2 Describe the pattern using variables and expressions. This rule tells how the letter grows.

3 Use the rule to predict the number of tiles needed for each *I*:

    a. size 12   **b.** size 15   **c.** size 22   **d.** size 100

Suppose you had 39 tiles. What is the largest size of *I* that you could make?

Using Variables and Expressions to Describe Patterns

These are the first three sizes of the letter *O*.

Size 1   Size 2   Size 3

How many tiles are needed to make each size? The pattern that tells how many can be described in words: The number of tiles needed is four times the size.

You can write this as 4 × *size*.

A shorter way to write the same thing is 4 × *s* or 4*s*.

In this example, the letter *s* is called a **variable** because it can take on many values.

4*s* is an **expression.** An expression is a combination of variables, numbers, and operations.

Relate the Rule to the Pattern

How can variables and expressions describe a pattern?

For one letter in the chart See How They Grow, the number of tiles is always $4s + 1$. The variable s stands for the size number. Decide how many tiles are added at each step. Look for the pattern.

1 Which letter do you think fits the pattern, *L*, *T*, or *X*? How many tiles are needed for size 16 of the letter?

2 For each of the other letters, give a rule that tells the number of tiles in any size.

Write About Your Own Letter Pattern

Make up your own letter shape with tiles. Figure out how you can make the letter grow into larger sizes.

- Draw your letter and tell how it grows.

- Give the rule for your letter using variables and expressions.

- Show how you can use the rule to predict the number of tiles that it would take to build size 100 of your letter.

See How They Grow

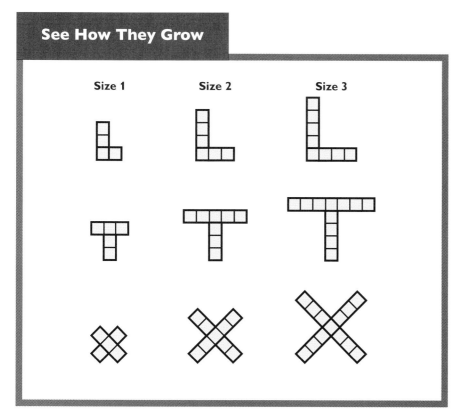

Size 1   Size 2   Size 3

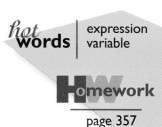

hot **words** | expression
variable

Homework

page 357

5 Tiling Garden Beds

DESCRIBING
PATTERNS WITH
EQUIVALENT
EXPRESSIONS

You have used tables and variables to describe different kinds of patterns. Here you will apply what you have learned to a new situation. You will show how each part of your solution relates to the situation. Then you can compare different ways of expressing the same idea.

Find the Number of Tiles

What different rules or expressions can you write to describe a pattern?

Here are three sizes of gardens framed with a single row of tiles:

Length 1   Length 2   Length 3

1 Begin a table that shows the number of tiles for each length. Use the table to write an expression that describes the number of tiles needed for a garden of any length.

Length of Garden	Number of Tiles
1	8
2	
3	

2 Use your expression to find how many tiles you would need to make a border around gardens of each of these lengths.

a. 20 squares   **b.** 30 squares   **c.** 100 squares

3 Tell how you would find the length of the garden if you knew only the number of tiles in the border.

Test your method. How long is the garden if the following numbers of tiles are used for the border?

a. 68 tiles   **b.** 152 tiles   **c.** 512 tiles

4 Relate each part of your expression to the garden and the tiles.

Extend the Rule

Some gardens are two squares wide, and vary in length. For example:

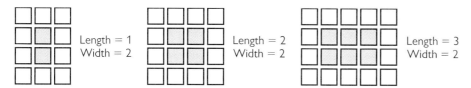

Length = 1
Width = 2

Length = 2
Width = 2

Length = 3
Width = 2

Can you figure out the number of tiles needed for gardens of any length and a width of 2? Use an expression that describes your method. Use the method to solve these problems.

How many tiles do you need to make a border around each of the following gardens?

1 $l = 5, w = 2$ **2** $l = 10, w = 2$

3 $l = 20, w = 2$ **4** $l = 100, w = 2$

> **Can you write a rule for tiling a garden of any length and any width?**

Write About Equivalent Expressions

You and your classmates may use different expressions to describe these patterns. Compare your ideas with others.

- What equivalent expressions did you and your classmates write?

- How do you know they are equivalent?

Conventions for Algebraic Notation

Writing variables and expressions in standard ways avoids confusion. The numeral is placed before the letter representing a variable:

$$2l \text{ not } l2$$

The numeral 1 is not required before a variable:

Use l instead of $1l$

"Two times the length" can be stated several ways:

$$2l \quad 2 \cdot l \quad 2 \times l$$

Place parentheses carefully.

$$l + 3 \times 2 \text{ does not equal } (l + 3) \times 2.$$

hot **words** | equivalent expression

omework

page 358

6 Chocolates by the Box

Some patterns can be described using more than one variable. The boxes of chocolates in this lesson are an example of this type of pattern. You will look for a way to write a rule that will apply to all sizes of chocolate boxes.

Find the Contents for Each Size

How can you use variables to describe the pattern?

Buy a Box of Chocolates—Get a Bonus

2 by 2 size

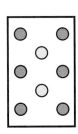

2 by 3 size

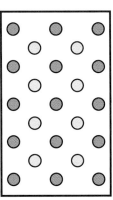

3 by 5 size

When you buy a box of Choco Chocolates, you get a bonus light chocolate between every group of four dark chocolates, as the diagrams show. The size of the box tells you how many columns and how many rows of dark chocolates come in the box.

1 How many *dark* chocolates will you get in each size of box?

a. 4 by 4	**b.** 4 by 8	**c.** 6 by 7
d. 12 by 25	**e.** 20 by 20	**f.** 100 by 100

2 How can you figure out the number of dark chocolates in any size box? Explain your method using words, diagrams, or expressions.

Rewrite Your Rule

Now that you have developed a rule for finding the number of dark chocolates, how can you figure out the number of light chocolates in any size box?

How can you use your method to write another rule?

1 Explain your method using words, diagrams, or expressions.

2 Test your method. How many *light* chocolates will you get in each box of these sizes?

a. 4 by 4   **b.** 4 by 8   **c.** 6 by 7

d. 12 by 25   **e.** 20 by 20   **f.** 100 by 100

3 Use variables and expressions to describe the total number of all chocolates in any box.

Write About Using Variables

Think about how you have used variables and expressions to describe patterns. Write a note to one of next year's students telling how you use these tools to solve problems.

- Give some examples of using expressions to describe patterns.

- Tell how you decide whether two expressions are equivalent.

A Pattern of Js with Two Variables

Choco Company uses chocolate squares to make letters you can eat. They make the letter *J* in different heights and widths.

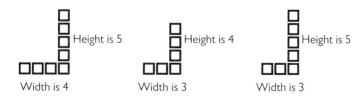

Height is 5   Height is 4   Height is 5
Width is 4   Width is 3   Width is 3

To find the total number of squares needed to make a letter *J*, you can add the height and width and subtract 1.

You can write this with words and symbols:
height + width − 1

Or you can use two variables (one for height and one for width) and say it this way: $h + w - 1$

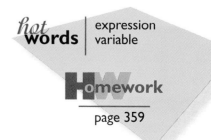

hot **words** | expression variable

Homework
page 359

PHASE THREE

In this phase, you will move around the coordinate plane to identify points and to graph data that you develop when you are looking for patterns. As you work with number rules for patterns, you begin to see the relationship of a list of ordered pairs to the line they form when they are graphed. You will use the information you find in the graphs to solve problems about some real-life situations involving summer jobs for two teenagers.

Describing Patterns Using Graphs

WHAT'S THE MATH?

Investigations in this section focus on:

PATTERN SEEKING

- Relating a pattern rule to a graphed line and the line to the rule

ALGEBRA

- Identifying points on the coordinate plane
- Making tables of ordered pairs
- Graphing ordered pairs on the coordinate plane

PROBLEM SOLVING

- Interpreting data from a graph
- Using patterns on a graph to solve problems

MathScape Online
mathscape1.com/self_check_quiz

 # Gridpoint Pictures

So far, you've described patterns using words, pictures, tables, variables, and expressions. Now you will look at another tool: the coordinate grid or plane. Before using this tool, review some of the basic definitions in "Finding Your Way Around the Coordinate Plane."

Describe a Picture on the Coordinate Plane

How can you make and describe a picture on a coordinate plane?

Draw a simple picture on a coordinate plane. You may want to write your initials in block letters or draw a house or other simple object. Be sure your picture has parts in all four quadrants.

Using coordinates, write a description of how to make your picture. You can also include other directions, but you may not use pictures.

Finding Your Way Around the Coordinate Plane

The coordinate plane is divided into four quadrants by the horizontal x-axis and the vertical y-axis. The axes intersect at the origin. You can locate any point on the plane if you know the coordinates for x and y. The x-coordinate is always stated first.

The y-axis

This is the first quadrant.

The x-axis

This is the point (−3, −2). The x-coordinate is −3. The y-coordinate is −2.

(0, 0) is the origin

Decode a Gridpoint Picture

When your picture and list are finished, trade descriptions with a partner. Do not show your pictures until later.

1 Draw the picture your partner has described. Keep notes on anything that is not clear.

2 Return the description and your picture to your partner. Check the picture your partner drew from your description to see if it matches your original drawing. If there are differences, what caused them? If necessary, revise your description.

How can coordinates help you draw a gridpoint picture?

Write About What You See

Suppose you are given the coordinates for a set of points. How could you tell if the points are all on a straight vertical line? a straight horizontal line? Explain.

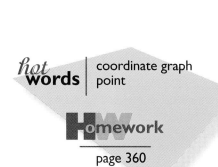

hot **words** | coordinate graph
point

Homework
page 360

8 Points, Plots, and Patterns

When you used points on a coordinate plane to describe a picture, you followed a visual pattern. Now you will see what happens when the points all spring from a number rule. Keep an eye out for patterns!

Find Some Patterns in the Plots

What patterns in the points on the coordinate grid fit a number rule?

You may think of a number rule as being expressed only in words and numerals. Some surprising things develop when you plot a graph using ordered pairs that follow a rule.

1 Make up a number rule of your own.

2 Make a table of points that fit the rule.

3 Plot the points on the coordinate plane.

4 What patterns do you notice on the coordinate plane? Can you find points that fit the rule but do not fit the pattern?

Repeat the process with other number rules. Keep a record of your results.

Ordered Pairs and Number Rules

In the ordered pair (4, 8), 4 is the *x*-coordinate and 8 is the *y*-coordinate.

A number rule tells how the two numbers in an ordered pair are related. Here are some examples:

- The *x*-coordinate is half the *y*-coordinate.
- The *y*-coordinate is 6 more than the *x*-coordinate.
- The *x*-coordinate and the *y*-coordinate are the same.

Find the Rule for the Pattern

Think about the patterns that you have been plotting on the grid. Do you think you can find the rule for a line passing through two points?

Try this. Mark the two endpoints that are given. Carefully draw a line between the points. What is the pattern or rule for all the points falling exactly on the line?

1 $(6, 4)$ and $(-6, -8)$

2 $(3, 12)$ and $(-1, -4)$

How does a line show a relationship between coordinates on a grid?

Write About Graphing Number Rules

What do you notice about the graphs from your number rules and those of your classmates? Write a summary of as many generalizations as possible. Here are some possibilities that you might want to include:

- What can you say about rules that give a line that passes through $(0, 0)$?

- How can you make parallel lines move up or down the graph?

- How can you change your rule to make a line steeper?

hot **words** | coordinate graph
ordered pair

HW**omework**

page 361

Payday at Planet Adventure

Rachel and Enrico have summer jobs at Planet Adventure, a local amusement park. Since Rachel gets bonus pay, it's not easy to make a quick comparison of their wages. You will use a graph to compare how much they earn for different lengths of time.

Make a Graph of Earnings

What can a graph tell about wages?

Rachel works in the Hall of Mirrors. Her rate of pay each day is $5 per hour. She also gets a daily $9 bonus for wearing a strange costume.

- Make a table to show what pay she should receive for different numbers of hours worked each day.

- Next, draw a graph of the data for Rachel's pay. Label the axes and choose an appropriate scale for the graph.

How would you find the amount of money Rachel earns for any number of hours worked? Use words, diagrams, or equations to explain your method.

Compare Earning Rates

Enrico works at the Space Shot roller coaster. His rate of pay is $6.50 an hour.

How can a graph help compare wages?

- Make a table to show what pay he receives for different numbers of hours worked each day.

- On the same grid you used for Rachel's pay, draw a graph of the data for Enrico's pay.

- How would you find the amount of pay Enrico earns for any number of hours worked? Is the method different from the one you used for Rachel?

Write About the Graphs

Compare the graphs of Rachel's pay and Enrico's pay. Which job pays better? How did you decide?

- Tell how the graphs are the same and how they are different.

- For what numbers of hours worked does Rachel earn more than Enrico? Enrico more than Rachel?

- Do Rachel and Enrico ever earn the same amount for the same number of hours worked? How did you find out?

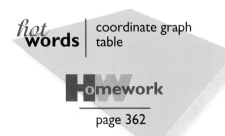

hot **words**   coordinate graph
table

Homework
page 362

PHASE FOUR

This phase gives you a chance to try out all the tools you have learned to use in describing patterns. You will examine the situations, decide how you will explain the pattern, and then write a rule to extend the pattern to any size. You will also use variables and expressions to describe patterns. You will soon see how much you have learned about looking for patterns.

Finding and Extending Patterns

WHAT'S THE MATH?

Investigations in this section focus on:

PATTERN SEEKING

- Identifying, describing, and generalizing patterns
- Choosing appropriate tools to describe patterns

NUMBER OPERATIONS

- Using inverse operations with pattern rules

ALGEBRA

- Using variables and expressions to describe patterns
- Making lists of ordered pairs to describe patterns
- Graphing ordered pairs

PROBLEM SOLVING

- Using patterns in problem situations

MathScape Online
mathscape1.com/self_check_quiz

10 Sneaking Up the Line

Finding a pattern in a simpler problem can help you understand a problem with greater values. Careful reasoning in a sample problem will help you make a rule about this sneaky situation.

Solve a Simpler Problem

How can you identify a pattern by solving a simpler problem first?

After you read "A Woolly Tale" and make your prediction, try some small problems to help look for a pattern.

1 Can you find how many sheep would be shorn before Eric if there are 6 sheep ahead of him? Use counters, diagrams, or any other method to solve the problem.

What if there are 11 sheep ahead of him? What if the number in front of Eric is 4 to 10? 11 to 13? It may help to make a table, then graph the data.

Use what you learned in the simpler problems.

2 Find how many sheep would be shorn before Eric if there were 49 sheep in front of him. Does the answer match the prediction you made at first?

3 Describe a rule or expression you would use to find the number shorn before Eric for any number of sheep in front of him.

A Woolly Tale

Eric the Sheep is at the end of a line of sheep waiting to be shorn. But being an impatient sort of a sheep, every time the shearer takes a sheep from the front to be shorn, Eric sneaks up the line two places.

Think about how long it will take Eric to reach the head of the line. Before you begin to work, make a prediction. If there are 49 sheep ahead of him, how many of the sheep will be shorn before Eric?

Test Your Rule

In each case, find how many sheep are shorn before Eric.

1. There were 37 sheep in front of Eric.

2. There were 296 sheep in front of him.

3. There were 1,000 sheep in front of him.

4. There were 7,695 sheep in front of the sneaky sheep.

Now try using your rule to find how many sheep were lined up in front of Eric if:

5. 13 sheep were shorn before him.

6. 21 sheep were shorn before Eric.

Will your rule work for any number?

Write About Your Sneaky Rule

Eric's pattern of sneaking up the line follows a rule with some new things to think about.

- How do situations 5 and 6 above differ from 1–4?

- Describe how you used your rule to find how many sheep were ahead of Eric.

hot **words** | coordinate graph
table

Homework

page 363

Something Fishy

EXPLORING
GEOMETRIC
PATTERNS OF
GROWTH

You have learned to identify the rule for many patterns from simple to complex. Here you find how to describe the pattern as an animal drawing changes in more than one way. Then you get to grow your own animals and describe their patterns.

Go Fishing for Some New Patterns of Growth

How can you describe the growth of some geometric patterns?

The fish in Patternville grow in a particular way. The diagram on dot paper shows the first four stages of this growth.

1 For fish in growth stages 1–6, figure out how many line segments and spots they would have in each stage. Show your answers using expressions, tables, and graphs.

2 Use what you observed about the growth in the number of *line segments* to answer these questions:

   a. How many line segments would a fish in stage 20 have?

   b. How many line segments would a fish in stage 101 have?

   c. In what stage of growth has the fish 98 line segments?

   d. In what stage of growth has the fish 399 line segments?

Something Fishy in Patternville

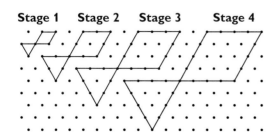

Stage 1   Stage 2   Stage 3   Stage 4

There are two ways to measure the growth of the fish: line segments and spots.

Line Segments
Count the line segments between dots needed to make the fish. It takes seven line segments to make the fish in stage 1.

Spots
Count the number of spots inside the fish's body (not including the tail). A fish in stage 1 has no spots and a fish in stage 2 has one spot.

3 Use what you observed about the growth in the number of *spots* to answer these questions:

    a. How many spots would a fish have in stage 20?

    b. How many spots would a fish have in stage 101?

How can you figure out how many line segments and spots a fish would have at any stage? Use words, diagrams, or equations to explain your method. Look for a short way to state it clearly.

Create Your Own Growing Patterns

Look at the examples from the Patternville Zoo. Then use graph paper, dot paper, or toothpicks to create your own "animal" and show how it grows in stages. Make sure you have a clear rule for the way it grows.

Draw your animal and write the rule for its growth at any stage on the back of the drawing. Create a table and a graph to show stages 1–5 of your animal's growth.

Write an explanation of why your rule will predict the number of dots or line segments for any size of your animal.

What are some different patterns that describe how a drawing can grow?

Examples from the Patternville Zoo

Stage	1	2	3
Dots	14	16	18
Area	6	7	8

Stage	1	2	3
Hexagons	1	4	7
Perimeter	6	18	30

Stage	1	2	3
Perimeter	11	22	33
Area in triangles	11	44	99

hot **words** | expression variable

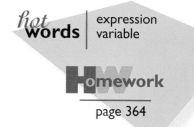

Homework
page 364

12

The Will

The results of a growth pattern are not always obvious at first. In this situation you will project the patterns into the future in order to make a good recommendation for Harriet. You may choose any tools you wish to describe the patterns.

Make a Prediction

Which payment plan looks like the best choice?

Harriet's uncle has just died. He has left her some money in his will, but she must decide how it is paid. Read about the three plans in the will. You can help her choose which plan would be the best for her.

Before doing any calculations, predict which plan would give Harriet the greatest amount of money at the end of year 25. Give your reasons for your choice.

From Harriet's Uncle's Will

... and to my niece Harriet, I give a cash amount of money to spend as she pleases at the end of each year for 25 years. Knowing how much she likes a bit of mathematics, I give her a choice of three payment plans.

Plan A:

$100	at the end of year 1
$300	at the end of year 2
$500	at the end of year 3
$700	at the end of year 4
$900	at the end of year 5
(and so on)	

Plan B:

$10	at the end of year 1
$40	at the end of year 2
$90	at the end of year 3
$160	at the end of year 4
$250	at the end of year 5
(and so on)	

Plan C:

1 cent	at the end of year 1
2 cents	at the end of year 2
4 cents	at the end of year 3
8 cents	at the end of year 4
16 cents	at the end of year 5
(and so on)	

Compare the Plans

How can you describe the patterns and extend them into the future?

Each plan pays a greater amount of money each year, but the amounts increase in different ways. In order to compare the plans, you will need to find the amount Harriet would receive for each plan in each year.

1 For each plan, what patterns did you see in the amount Harriet would receive each year? Use words, diagrams, or expressions to explain the patterns.

2 Make tables and graphs to show the amount she would receive for each plan in each year. What amount of money will Harriet receive at the end of the tenth year? the twentieth year? the twenty-fifth year?

3 Describe a method you could use to find the amount Harriet would get at the end of any year for each plan.

Give Some Good Advice

Write a letter to Harriet advising her as to which plan she should choose and why. Make sure to compare the three plans. You can choose words, diagrams, tables, graphs, and expressions to support your recommendation.

hot **words** | coordinate graph | expression

Homework
page 365

Calendar Tricks

Applying Skills

Read the four statements A–D. Then tell which are true for each two-by-two block of four numbers.

A. The sum of the four numbers is divisible by 4.

B. The sum of the four numbers is divisible by 8.

C. The sum of the bottom two numbers differs from the sum of the top two numbers by 14.

D. The number in the bottom right corner is three times as big as the number in the top left corner.

Sun	Mon	Tue	Wed	Thurs	Fri	Sat
1	2	3	4	5	6	7
8	9	10	11	12	13	14
15	16	17	18	19	20	21
22	23	24	25	26	27	28
29	30	31				

1. Which statements are true for the shaded block on the calendar?

2. Which statements are true for the block formed by numbers 6, 7, 13, and 14?

3. Which statements are true for the block formed by numbers 19, 20, 26, and 27?

4. Which statements do you think are true for every possible two-by-two block of four numbers on the calendar?

Extending Concepts

5. Choose two different blocks of nine numbers arranged 3 across and 3 down on a calendar, and check that this rule holds: For any block of nine numbers the average of the four corner numbers is equal to the middle number. Show your work and explain why the rule works.

6. Make up your own trick which works for any three-by-three block of nine numbers. Explain why your trick works.

Making Connections

7. It is said that the seven-day week was based originally on the idea of the influence of the planets. For a long time people believed that seven celestial bodies revolved around the earth. The early Romans observed an eight-day week based on the recurrence of market days.

a. What pattern or trick do you notice about numbers on a diagonal of a calendar such as 2, 10, 18,…? Why does this trick work? Revise the rule for the pattern so that it would work for a calendar with 8 days in each row.

b. Would the rule in item **5** work for a calendar with 8 days in each row? Why or why not?

Painting Faces

Applying Skills

A company that makes colored rods uses a paint stamping block to paint only the front and one end of each rod like this:

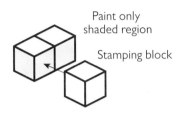

Paint only
shaded region

Stamping block

1. Copy and complete the table to show how many stamps would be needed to paint rods with lengths 1 to 10.

Length of Rod	Stamps Needed
1	2
2	3
3	
4	
5	
6	
7	
8	
9	
10	

2. What rule could you use to find the number of stamps needed for a rod of any length?

3. How many paint stamps are needed to paint a rod of length 23? 36? 64?

4. How long is the rod if the number of stamps needed is 23? 55? 217?

Extending Concepts

Suppose the company also makes cubes of different sizes and uses the stamping block to paint only the front face of each cube as shown.

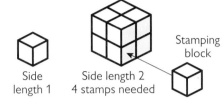

Side
length 1

Side length 2
4 stamps needed

Stamping
block

5. Make a table to show the number of paint stamps needed for cubes with side lengths 1 to 6. What pattern do you notice? Write a general rule for finding the number of paint stamps needed for any cube.

6. Use your rule from item **5** to find how many stamps you would need for a cube with side length 43.

Writing

7. Answer the letter to Dr. Math.

> Dear Dr. Math,
>
> I decided I didn't need to figure out the number of stamps that would completely cover every different size rod. Instead, I just tested some sample lengths and made a table like this:
>
> | Length of Rod | Stamps Needed |
> |:---:|:---:|
> | 1 | 6 |
> | 3 | 14 |
> | 5 | 22 |
> | 10 | 42 |
> | 20 | 82 |
> | 100 | 402 |
>
> But I'm confused and can't see a pattern. Was this a good shortcut?
> Pat Turn

Crossing the River

Applying Skills

Suppose that a group of hikers with exactly two children and a number of adults must cross a river in a small boat. The boat can hold either one adult, one child, or two children. Anyone can row the boat.

How many one-way trips are needed to get everyone to the other side if the number of adults is:

1. 3?   **2.** 4?   **3.** 5?

4. Make a table that records the number of one-way trips needed for all numbers of adults from 1 to 8.

How many trips are needed if the number of adults is:

5. 11?   **6.** 37?

7. 93?   **8.** 124?

How many adults are in the group if the number of trips needed is:

9. 25?   **10.** 49?

11. 73?   **12.** 561?

Extending Concepts

13. Find the number of one-way trips that would be needed for a group of 4 adults and 3 children to cross the river. Use a recording method that keeps a running total of the number of trips and the number of adults and children on each side of the river.

14. Make a table showing the number of trips needed for 1 to 5 adults and 3 children to cross the river. Describe in words a general rule that you could use to find the number of trips needed for any number of adults and 3 children. How did your table help you to find the rule?

15. How many trips would it take for 82 adults and 3 children to cross?

16. How many adults are in the group if 47 trips are needed to get all the adults and 3 children across the river?

Making Connections

17. For transportation, the Hupa Indians of Northwestern California used canoes hollowed out of half of a redwood log. These canoes could carry up to 5 adults. If one adult could row the canoe, how many one-way trips would be needed for 17 adults to cross a river using one of these canoes?

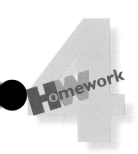

Letter Perfect

Applying Skills

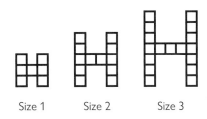

Size 1   Size 2   Size 3

1. Draw an *H* of size 4 and an *H* of size 5.

2. How many tiles are needed for each of the sizes 1 to 5 of the letter *H*?

3. How many tiles are added at each step to the letter *H*?

4. Which of these expressions tells how many tiles are needed for a letter *H*? The variable *s* stands for the size number.

$2s$ $5s$ $2s + 5$ $5s + 2$ $4s + 3$

5. Predict the number of tiles needed for a letter *H* of these sizes:

   a. 11   **b.** 19   **c.** 28

   d. 57   **e.** 129

6. What is the largest size of *H* you could make if you had:

   a. 42 tiles?   **b.** 52 tiles?

   c. 127 tiles?   **d.** 152 tiles?

Extending Concepts

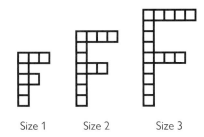

Size 1   Size 2   Size 3

7. a. Find how many tiles are needed for a letter *F* of sizes 1 to 5. Then describe in words a rule you could use to find the number of tiles needed for a letter *F* of any size.

   b. Write an expression to describe your rule in item **7a.** Use the letter *s* to stand for the size number.

   c. Explain why *s* is called a variable.

8. For a mystery letter, the number of tiles needed is $3s + 8$. The variable *s* stands for the size number. How many tiles are added at each step? How can you tell? Which letter grows faster, the mystery letter or the letter *F*?

Making Connections

9. The Celsius temperature scale uses the freezing point of water as 0 degrees Celsius and its boiling point as 100 degrees Celsius. If *c* stands for a known Celsius temperature, the expression $1.8c + 32$ can be used to find the Fahrenheit temperature. What Fahrenheit temperature corresponds to 16 degrees Celsius? to 25 degrees Celsius? How did you find your answers?

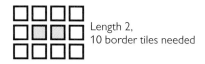

Tiling Garden Beds

Applying Skills

Find the value of these expressions if $l = 3$ and $w = 5$:

1. $6l$ **2.** $3l + w$ **3.** $4(l + w)$

4. $lw - 2$ **5.** $l(w - 2)$

For items **6–17**, assume that each garden is one square wide. Suppose you want to frame your garden with a single row of tiles like this:

```
Length 2,
10 border tiles needed
```

Find the number of border tiles needed for a garden of each length.

6. 4   **7.** 7   **8.** 13

9. 25   **10.** 57   **11.** 186

Find the length of the garden if the number of border tiles needed is:

12. 16   **13.** 38   **14.** 100

15. 370   **16.** 606

17. Which of these expressions could *not* be used to find the number of tiles needed to make a border around a garden with length l squares?

a. $(l + 3) \times 2$ **b.** $(l + 2) \times 2 + 2$

c. $2l + 6$ **d.** $(l + 2) \times 2$

e. $(l + 1) \times 2 + 4$

Extending Concepts

18. a. Write two different expressions for finding the number of tiles needed to make a border around a garden whose width is 4 and whose length may vary. Use l to represent the length.

b. Explain why each of the expressions makes sense by relating each part of the expression to the garden and the tiles.

c. How many tiles are needed to make a border around a garden with a width of 4 and a length of 35?

19. Suppose that one length of the garden is along a wall like this:

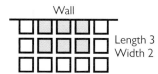

The length and the width may both vary. Write an expression for the number of tiles needed. Use l for the length and w for the width.

Making Connections

20. A Japanese garden is considered a place to contemplate nature. An enclosure such as a bamboo fence is often used to separate the garden from the everyday world outside and to create the feeling of a sanctuary. What length of fencing would be needed to enclose a rectangular garden 70 feet long and 30 feet wide?

Chocolates by the Box

Applying Skills

In a box of Choco Chocolates there is a light chocolate between each group of 4 dark chocolates as shown. The size of the box tells the number of columns and rows of dark chocolates.

3 by 4 size

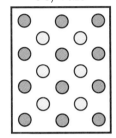

Find the number of dark chocolates and the number of light chocolates in boxes of these sizes:

1. 5 by 5   **2.** 8 by 10

3. 15 by 30   **4.** 40 by 75

Give the total number of chocolates in boxes of these sizes:

5. 3 by 5   **6.** 7 by 9

7. 18 by 20   **8.** 30 by 37

Extending Concepts

9. Either of the two equivalent expressions $lw + (l - 1) \times (w - 1)$ or $2lw - l - w + 1$ may be used to find the total number of chocolates in a box of Choco Chocolates.

a. Verify that both expressions give the same result for a 16 by 6 box.

b. Explain why the first expression makes sense.

10. Choco Chocolates wants to make new triangle-shaped boxes as shown. Write rules for finding the number of light chocolates, dark chocolates, and total chocolates in a triangular box of any size. How many dark and how many light chocolates are in a box of size 9? size 22?

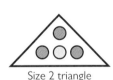

Size 2 triangle

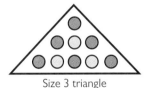

Size 3 triangle

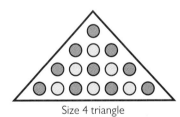

Size 4 triangle

Writing

11. Write a paragraph describing how you have used variables and expressions to describe patterns. Be sure to explain the meaning of the words *variable* and *expression*. Give some examples of using expressions to describe patterns.

Gridpoint Pictures

Applying Skills

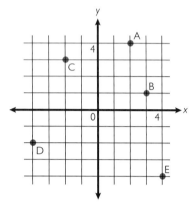

1. Give the coordinates of each of the five points shown on the coordinate plane.

2. Draw a grid and axes similar to those in item **1.** Place a dot at each point.

    a. $(1, 4)$ **b.** $(-3, 4)$ **c.** $(-1, 0)$

    d. $(-3, -2)$ **e.** $(1, -2)$

Tell whether each statement is true or false:

3. The y-coordinate of the point $(3, -5)$ is 3.

4. The x-coordinate of the point $(-9, 1)$ is -9.

5. The point $(0, -5)$ lies on the x-axis.

6. The point $(0, 4)$ lies on the y-axis.

7. The points $(2, 5)$ and $(-3, 5)$ lie on the same horizontal line.

Extending Concepts

8. Using coordinates, tell how to make this picture. You can include other directions but you may not use pictures.

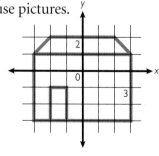

9. a. Give the coordinates of three points which lie on the x-axis. How can you tell if a point lies on the x-axis?

    b. Give the coordinates of two points which lie on the same vertical line. How can you tell if two points lie on the same vertical line?

Writing

10. Answer the letter to Dr. Math.

> Dear Dr. Math,
>
> Tom asked me to predict what I would get if I started at (3, −5) on a coordinate plane and drew a line to (3, 1), then another line to (3, 7). I noticed right away that all the points have the same x-coordinate. I know that the x-axis is horizontal, so I figured that I would get a horizontal line. Is this good reasoning? I have to know for sure because Tom will really gloat if I get this wrong.
>
> K. O. R. Denate

Points, Plots, and Patterns

Applying Skills

Rule A: The *y*-coordinate is three times the *x*-coordinate.

Rule B: The *y*-coordinate is three more than the *x*-coordinate.

Read Rules A and B, then tell which of the ordered pairs below satisfy each rule.

1. $(2, 6)$ **2.** $(12, 4)$ **3.** $(3, 6)$

4. $(0, 3)$ **5.** $(-1, -3)$ **6.** $(5, 2)$

Read Rules C, D, and E, then answer items **7** and **8.**

Rule C: The *y*-coordinate is twice the *x*-coordinate.

Rule D: The *y*-coordinate is six more than the *x*-coordinate.

Rule E: The *y*-coordinate is five times the *x*-coordinate.

7. Of C, D, and E, which rule or rules produce a line passing through the origin?

8. Of C, D, and E, which rule produces the steepest line?

9. a. Copy and complete this table using **Rule F:** The *x*-coordinate is two less than the *y*-coordinate.

x	*y*
	4
	-2
3	
-1	

b. Plot the points from your table on a coordinate grid. Do the points lie on a straight line?

Extending Concepts

10. Make a table of points that fit this rule: The *y*-coordinate is twice the *x*-coordinate. Plot the points on a coordinate plane and draw a line through the points.

11. Pick a new point on the line you plotted in item 10. What do you notice about its coordinates? Do you think that this would be true for any point on the line?

12. Make up two different number rules which would produce two parallel lines.

13. Make up a number rule which would produce a line that slopes downward from left to right.

Making Connections

In a *polygon,* the number of diagonals that can be drawn from one *vertex* is 3 fewer than the number of sides of the polygon.

6 sides, 3 diagonals

In the table below, *x* represents the number of sides of a polygon and *y* represents the number of diagonals that can be drawn from one vertex.

14. Complete the table and plot the points on a coordinate plane. What do you notice?

x	*y*
5	2
6	
9	
	8

Payday at Planet Adventure

Applying Skills

At Planet Adventure, Lisa works at the waterslide. Her rate of pay is $9 per hour. Joel works at the Mystery Ride and makes $5 per hour plus a one-time $12 bonus for dressing up in a clown costume.

1. Make a table to show the pay Lisa would receive for 1–7 hours of work.

2. How many hours would Lisa have to work to earn:
   a. $81? **b.** $99?
   c. $31.50? **d.** $58.50?

3. Make a table to show the pay Joel would receive for 1–7 hours of work.

4. How much will Joel earn if he works 8 hours? 4.5 hours? 7.5 hours?

5. How many hours would Joel have to work to earn:
   a. $57? **b.** $67?
   c. $39.50? **d.** $54.50?

6. Copy the axes shown. Make a graph of the data for Lisa's pay. On the same grid, make a graph of the data for Joel's pay.

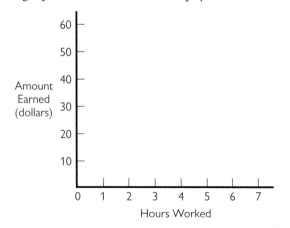

Extending Concepts

7. Write a rule that you could use to find the wage earned by Joel for any number of hours of work.

8. For what number of hours of work will Lisa and Joel earn the same amount? How can you figure this out using the graphs? without using the graphs?

9. For what number of hours worked will Joel earn more than Lisa?

10. Why is Lisa's graph steeper than Joel's?

11. Why does Joel's graph not pass through the origin?

12. If Joel works 4 hours, how much does he earn? What is his rate *per hour*? If he works more than 4 hours, is his average rate of income per hour higher or lower than this?

Writing

13. Answer the letter to Dr. Math.

Dear Dr. Math,

I've been offered two different jobs. The first one would pay $8 per hour. The second one would pay a fixed $24 each day plus $5 per hour. Which job should I take if I want to earn as much money as possible? How do I figure it out?

Broke in Brokleton

Sneaking Up the Line

Applying Skills

Eric the Sheep is waiting in line to be shorn. Each time a sheep at the front of the line gets shorn, Eric sneaks up the line two places.

How many sheep will be shorn before Eric if the number of sheep in front of him is:

1. 5?   **2.** 8?

3. 15?   **4.** 42?

5. 112?   **6.** 572?

How many sheep could have been in line in front of Eric if the number of sheep shorn before him is:

7. 7?   **8.** 12?

9. 25?   **10.** 39?

Extending Concepts

For the following questions, suppose that each time a sheep at the front of the line gets shorn, Eric sneaks up *three* places instead of two.

11. Complete the table for this situation. Then graph the data.

Number of Sheep In Front of Eric	Number of Sheep Shorn Before Eric
4	
5	
6	
7	
8	
9	
10	
11	

12. How many sheep will be shorn before Eric if there are 66 sheep in front of him? How did you figure this out?

13. How could you find the number of sheep shorn before Eric for *any* number of sheep in front of him? Describe two different methods you could use and explain why each method works.

14. If 28 sheep are shorn before Eric, how many sheep could have been in front of him? How did you figure it out? Why is the answer not unique?

Writing

15. Describe a method you could use to find the number of sheep shorn before Eric for any number of sheep in front of him and for any number of sheep that Eric sneaks past.

PATTERNS IN NUMBERS AND SHAPES • HOMEWORK 10   **363**

Something Fishy

Applying Skills

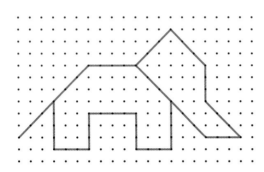

This elephant drawing in growth stage 3 has a height of 10 units, a length of 19 units, and a body area (excluding the head) of 49 square units.

1. Figure out what the height, length, and body area would be for elephants in growth stages 1−6. Show your answers in a table.

2. Make graphs to show the height, length, and body area for elephant drawings in growth stages 1−6.

Use the patterns that you observe in your table to find the height, length, and body area for elephant drawings in each of these growth stages:

3. 12   **4.** 20

5. 72   **6.** 103

7. In what stage of growth is the elephant if its height is 28? 49? 76? 211?

8. In what stage of growth is the elephant if its length is 67? 133? 325?

Extending Concepts

9. Use words and expressions to describe rules you could use to find the height, length, and body area for elephant drawings in any growth stage.

10. What growth stage is an elephant in if its body area is 529? Explain how you figured out your answer.

11. Create your own "animal" and draw pictures to show how it grows in stages. Make sure you have a clear rule for the way it grows. Make a table to show stages 1−5 of your animal's growth. Describe in words a rule for its growth. Describe your rule using variables and expressions.

Writing

12. Do you think that in reality, the height of an actual elephant is likely to increase according to the pattern you described in item **9**? Why or why not?

The Will

Applying Skills

Annie may receive money according to any one of three plans. The amounts of money that each plan would yield at the end of years 1, 2, 3, 4, and 5, respectively, are as follows:

Plan A: $100, $250, $400, $550, $700, and so on.

Plan B: $10, $40, $100, $190, $310, and so on.

Plan C: 1 cent, 3 cents, 9 cents, 27 cents, 81 cents, and so on.

1. Make a table showing the amount of money Annie would receive from each plan at the end of years 1–15.

2. For each plan, make a graph showing the amount of money Annie would receive at the end of years 1–10.

3. Which plan yields the most money at the end of year 8? 12? 17? 20?

4. Describe in words the growth pattern for each plan.

Extending Concepts

5. At the end of which year will Annie first receive more than $5,000 if she uses Plan A? Plan B? Plan C?

6. Which plan would you recommend to Annie? Why?

7. Figure out a pattern which gives more money at the end of the fifteenth year than Plan A but not as much as Plan B.

Making Connections

8. The **half-life** of a radioactive substance is the time required for one-half of any given amount of the substance to decay. Half-lives can be used to date events from the Earth's past. Uranium has a half-life of 4.5 billion years! Suppose that the half-life of a particular substance is 6 days and that 400 grams are present initially. Then the amount remaining will be 200 grams after 6 days, 100 grams after 12 days, and so on.

   a. What amount will remain after 18 days? after 24 days?

   b. When will the amount remaining reach 3.125 grams?

   c. Will the amount remaining ever reach zero? Why or why not?

Glencoe

This unit of MathScape: Seeing and Thinking Mathematically was developed by the Seeing and Thinking Mathematically project (STM), based at Education Development Center, Inc. (EDC), a non-profit educational research and development organization in Newton, MA. The STM project was supported, in part, by the National Science Foundation Grant No. 9054677. Opinions expressed are those of the authors and not necessarily those of the Foundation.

CREDITS: Unless otherwise indicated below, all photography by Chris Conroy.

321 (tc)Bruce Stromberg/Graphistock, (tr)Jeremy Walker/Getty Images; **338–339** Bruce Stromberg/Graphistock; **345** (b)Chad Ehlers/Photo Network; **346–347** Jeremy Walker/Getty Images.

The **McGraw-Hill** Companies

Send all inquiries to:
Glencoe/McGraw-Hill
8787 Orion Place
Columbus, OH 43240-4027

ISBN: 0-07-866804-2

4 5 6 7 8 9 10 058 07 06